LIFE SCIENCE IN DEPTH

CELLS
AND
CELL FUNCTION

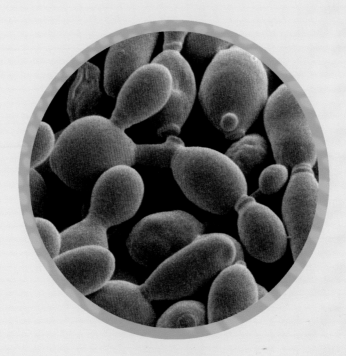

Sally Morgan

Heinemann
LIBRARY

www.heinemann.co.uk/library
Visit our website to find out more information about Heinemann Library books.

To order:

 Phone 44 (0) 1865 888066

Send a fax to 44 (0) 1865 314091

 Visit the Heinemann bookshop at www.heinemann.co.uk/library to browse our catalogue and order online.

First published in Great Britain by
Heinemann Library, Halley Court, Jordan Hill,
Oxford OX2 8EJ, part of Harcourt Education.

Heinemann is a registered trademark of
Harcourt Education Ltd.

Editorial: Sarah Shannon and Dave Harris
Design: Richard Parker and Q2A Solutions
Illustrations: Q2A Solutions
Picture Research: Natalie Gray
Production: Chloe Bloom

Originated by Modern Age Repro
Printed and bound in China by South China
Printing Company

10 digit ISBN: 0 431 10896 X
13 digit ISBN: 978 0 431 10896 4

10 09 08 07 06
10 9 8 7 6 5 4 3 2 1

British Library Cataloguing in Publication Data
Morgan, Sally
 Cells and cell function.
 - (Life science in depth)
 571.6
A full catalogue record for this book is available
from the British Library.

Acknowledgements
The publishers would like to thank the following
for permission to reproduce photographs:
Alamy pp. **9** (Bob Schuchman), **56** (Crispin
Hughes); BBC p. **59**; Corbis pp. **14** (Neil
Miller), **21**; Getty Images p. **5** (Bob Thomas);
Oxford Scientific pp. **28, 48**; Science Photo
Library pp. **57** (Alfred Pasieka), **47** (Astrid &
Hanns-Frieder Michler), **22** (Biology Media),
7 (BSIP/Sercomi), **54** (CNRI), **1, 38** (David
Scharf), **24** (Dr Yorgos Nikas), **20, 31**
(J.C.Revy), **52** (Kent Wood), **25** (M.I. Walker),
51 (Rosenfield Images Ltd), **43** (Simon Fraser),
19, 32, 44.

Cover photograph of a cell, reproduced with
permission of Science Photo Library/David Mack.

Our thanks to Emma Leatherbarrow for her
assistance in the preparation of this book.

Every effort has been made to contact copyright
holders of any material reproduced in this book.
Any omissions will be rectified in subsequent
printings if notice is given to the publishers.
The paper used to print this book comes from
sustainable resources.

Contents

Words printed in the text in bold, **like this**,
are explained in the Glossary.

What are cells?

The cell is the basic building block of life. When you look at yourself in a mirror, you see trillions of cells all working together. The smallest **organisms**, such as **bacteria**, consist of just a single cell, but most animals, plants, and **fungi** are multicellular, which means they are made up of many cells.

LOTS OF DIFFERENT TYPES OF CELL

Not only is the human body made up of trillions of cells, but there are about 200 different types of cell. Each type has a specific job or role to carry out in the body. Our muscles are made of muscle cells, our livers of liver cells, and there are even very specialized types of cells that make the enamel of our teeth or the clear **lenses** in our eyes. Usually, the structure of these different types of cells is specially **adapted** to their function. Cells of the same type group together to form **tissues**, and tissues form **organs**.

LARGE AND SMALL

Some cells are large and can be seen with the eye. The largest cell is the unfertilized egg cell of the ostrich and it is a massive 15 centimetres (6 inches) long! This is unusual since most cells are just a fraction of a millimetre in length and can only be seen with a microscope. The largest human cells are about the diameter of a human hair, but most are smaller, about one-tenth of the diameter of a human hair.

GROWING MORE

One of the characteristics of life is growth. Growth occurs when cells get larger and divide to form new cells. Human beings begin life as a single cell – a fertilized egg. This cell divides into two and then four, eight, sixteen, and so on.

The individual cells grow to a maximum size and then they divide. As you grow, you get more cells, so an adult has more cells than a child.

This book looks at the many different types of cells that are found in animals and plants as well as bacteria and **viruses**. It examines why most cells, whether from an elephant or an earthworm, are microscopic in size. It also answers questions such as: Why don't cells grow larger and larger to become giant cells? How can a cell divide to form two new cells? This book also covers some of the issues relating to cells today, such as **cloning** and **gene technology**.

The size of cells does not vary between people. A large person does not have larger cells than a smaller person, just more of them.

Inside a cell

Most cells are so small that they cannot be seen by the human eye. The only way to study them is to look at them using a microscope. This way they can be viewed hundreds, even thousands of times their normal size. There are two types of microscope – the light microscope and the **electron** microscope.

THE LIGHT MICROSCOPE

Microscopes work in a similar way to a magnifying glass. A magnifying glass uses a lens to **focus** the light reflecting off an object into a larger image. In a light microscope, there is an eyepiece lens and up to three objective lenses. The specimen is placed on the stage under one of the objective lenses. It is illuminated by a light placed under the stage. The objective lens is either raised or lowered to focus on the specimen. When the user looks through the eyepiece they see a magnified image of the specimen.

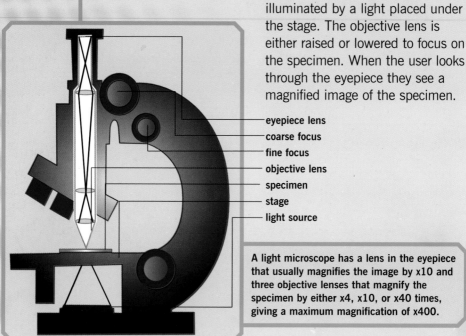

- eyepiece lens
- coarse focus
- fine focus
- objective lens
- specimen
- stage
- light source

A light microscope has a lens in the eyepiece that usually magnifies the image by x10 and three objective lenses that magnify the specimen by either x4, x10, or x40 times, giving a maximum magnification of x400.

UNDERSTANDING UNITS

When using microscopes, it is important to remember that the size of the different structures is really small. This table may help you work out just how small something is.

1 millimetre (mm) = 0.001 metre (m)

1 micrometre (μm) = 0.000001 metre (m)

1 nanometre (nm) = 0.000000001 metre (m)

ELECTRON MICROSCOPES

Most light microscopes give a maximum magnification of 400 times. The best light microscopes can magnify something about 2,000 times its original size, but an electron microscope can magnify a specimen up to two million times, allowing scientists to see cells in incredible detail. Electron microscopes, as the name suggests, use electrons rather than light to see the specimen. There are two common types of electron microscopes: transmission and scanning. The transmission electron microscope works in much the same way as a light microscope. These microscopes are used to study the contents of cells. A scanning electron microscope produces a picture of the surface of the specimen. It is useful for looking at the surface of cells, for example. In both types of electron microscope, the image is displayed on a monitor.

nucleus —

This image of an animal cell was taken using a transmission electron microscope. It is possible to see the tiny structures in the cell, such as the nucleus.

LOOKING AT CELLS

When a cell is viewed using a light microscope, it is possible to see that the cell is surrounded by a thin layer. This is called the **cell membrane** and it stops the contents of the cell from escaping. The cell membrane controls the passage of materials into and out of the cell. Normally it allows small **molecules** such as water, oxygen, and carbon dioxide to pass through but blocks the passage of large molecules such as **proteins**. These large molecules have to be moved across the membrane using special carriers.

Materials can pass across a cell membrane in different ways. Small molecules such as oxygen **diffuse** across the membrane. Diffusion is the movement of molecules from where they are in high concentration to where they are in low concentration. This means that oxygen diffuses into a cell because there is a lower concentration inside the cell than outside. Water moves across the membrane by a special diffusion process called **osmosis**. Large molecules such as proteins need to be carried

A typical animal cell is rounded in shape, surrounded by a cell membrane. This holds in the cytoplasm, which contains other structures such as the nucleus, ribosomes, and mitochondria.

- cell membrane
- cytoplasm
- nucleus
- endoplasmic reticulum
- Golgi body
- free ribosome
- ribosome
- mitochondrion
- lysosome

SCIENCE PIONEERS Robert Hooke

Robert Hooke built one of the earliest microscopes using a series of lenses which he used to look at living material. He examined tiny slivers of cork and described it as being made up of lots of tiny boxes that he called cells. His work changed the way scientists viewed the structure of living things. However, it was another 168 years before scientists observed the contents of cells.

across the membrane. Carrier molecules in the membrane pick up proteins on one side of the membrane and move them into the cell. This is called active transport. Unlike diffusion and osmosis, active transport needs energy to take place.

The **cytoplasm** fills the cell. It is made up of mostly water, often about 70 percent, the rest being molecules of salts, sugars, fats, **amino acids**, and proteins. Some of the proteins are **enzymes** that the cell has made. Enzymes control the rate of chemical reactions within the cell.

The largest structure in the cell is the **nucleus**. It usually lies towards the middle of the cell and is the most important part of the cell. It is the control centre and sends out messages to other parts of the cell, controlling the production of molecules such as proteins. The ability to control production comes from the **chromosomes**, which lie within the nucleus. Chromosomes can only be seen clearly at the time when the cell is dividing, when they appear as threads within the nucleus. In human cells there are 46 chromosomes or 23 pairs. Chromosomes are made from **DNA (deoxyribonucleic acid)** and protein. The DNA holds the genetic code that is used to make proteins.

The difference in size between an egg (female sex cell) and a sperm (male sex cell) can be seen clearly here as the sperm cluster around the egg.

PLANT CELLS

Plant cells differ from animal cells in several ways. One obvious difference is their size – plant cells tend to be larger than animal cells. They also have a regular shape because their cell membrane is surrounded by a cell wall that is rigid and gives the cell a fixed shape. The cell wall contains a substance called **cellulose**. Cellulose is made up of lots of long chains of **glucose**. The wall is strong but **permeable** so gases and water can pass through. Due to their regular shape, plant cells can pack tightly together leaving no air spaces. Some plant cells are more polygonal (five-sided) in shape and often they have extra cellulose laid down in the corners to provide additional support which is needed in stems.

The main features in a plant cell include the cellulose cell wall, the large central vacuole, and chloroplasts in the cytoplasm.

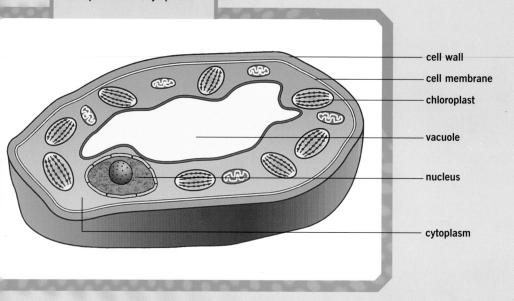

- cell wall
- cell membrane
- chloroplast
- vacuole
- nucleus
- cytoplasm

THE VACUOLE AND SUPPORT

Plant cells have a large space in the centre called a vacuole. This contains cell sap, which is a watery solution of sugar and salts. The vacuole is so large that it pushes the cytoplasm and nucleus to the side. The vacuole pushes on the contents, making the whole cell very firm. This helps to support the plant. If a plant does not get enough water, it wilts and its shoots droop. This happens because the vacuole loses water and the cells shrink slightly. This means there is less support so the leaf or the stem loses rigidity.

THE CYTOPLASM AND PHOTOSYNTHESIS

The cytoplasm of a plant cell usually contains small particles that contain **pigments**, **starch**, or oil. These particles are called **plastids**. **Chloroplasts** contain the green pigment **chlorophyll** and they are involved in the process of **photosynthesis**. A chloroplast is usually disc-shaped and about 5–8 μm in length and 2–4 μm wide. Cells where photosynthesis happens have between 20 and 40 chloroplasts. Cells in the root contain colourless plastids that are full of starch. The starch is used as a food store by the plant. For example, potato **tubers** contain cells that are full of starch grains.

PLANT OR ANIMAL?

The differences between plant and animal cells are shown in this table:

Plant cells	Animal cells
Cellulose cell wall outside of the cell membrane	No cell wall
Many plant cells have a regular shape	Usually irregular in shape
Many plant cells have a large vacuole filled with cell sap	Only have tiny vacuoles
Often have chloroplasts which contain chlorophyll	No chloroplasts
Some plant cells have starch grains	No starch grains

JOURNEY INTO THE CELL

The electron microscope has allowed scientists to really explore the cell. They have discovered a "cellular city" of tiny structures called **organelles**. Each type of organelle performs a particular function. For example, there are power stations (mitochondria), waste disposal stations (lysosomes), protein factories (ribosomes), and protein packaging and export facilities (the endoplasmic reticulum and Golgi body).

This diagram shows the size of different structures such as cells, organelles, bacteria, and viruses. It also shows the different sizes that are visible using either an electron microscope or a light microscope.

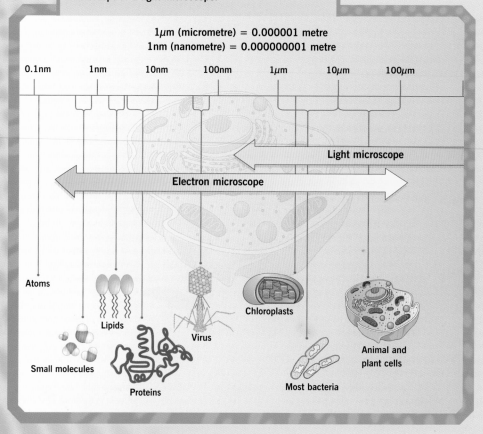

$1\mu m$ (micrometre) $= 0.000001$ metre
$1nm$ (nanometre) $= 0.000000001$ metre

0.1nm   1nm   10nm   100nm $1\mu m$ $10\mu m$ $100\mu m$

Light microscope

Electron microscope

Atoms

Lipids

Small molecules

Proteins

Virus

Chloroplasts

Most bacteria

Animal and plant cells

POWER CELLS

The mitochondria (singular mitochondrion) are the power stations of the cells. They are the sites of **aerobic respiration** and produce nearly all the energy the cells require to function. The energy is locked up in a very useful molecule called ATP – all cells use this for energy. Mitochondria have two membranes, an inner and an outer membrane. The inner membrane is highly folded and this increases its surface area for making ATP.

TRANSPORT AND MAKING PROTEINS

The endoplasmic reticulum (ER) is the transport network for the cell. It is made up of a network of tubes that criss-cross the cell. It allows substances such as protein to be moved from one part of the cell to another.

Some of the ER has tiny particles called ribosomes attached to it. Ribosomes are the protein factories where the manufacturing of protein takes place. This is called protein synthesis. Proteins are made up of lots of smaller molecules called amino acids. Amino acids are carried to the ribosomes and joined together in long chains to form proteins. Protein synthesis is extremely important to cells, so there are often hundreds or thousands of ribosomes found in cells.

MANUFACTURING AND WASTE DISPOSAL

The Golgi body looks a bit like a stack of tubes. It is responsible for making new products. For example, it makes glycoproteins from proteins and **carbohydrates**. It packages the new products into small membrane-bound sacs called vesicles which are moved around the cell or sent outside the cell.

The lysosomes are responsible for breaking down waste materials and old organelles within the cell. A lysosome is simply a bundle of enzymes contained within a membrane. It fuses with unwanted materials and pours its enzymes over them in order to break them down. The materials from this breakdown are either recycled in the cell or sent outside.

All sorts of cells

There are many different types of cells in existence, each one slightly different. There are some organisms that consist of just one large cell that does everything, while complex organisms such as mammals have many different types of cell, each adapted to carry out a particular job.

SPECIALIZED CELLS

The amoeba is a single-celled organism that carries out all its functions within its one cell. In a slightly more complex animal such as a hydra or jellyfish, there are seven types of cells, each with a particular job. There are sensory cells that react to touch, sting cells that are used by the hydra to catch and poison prey, and nerve cells that carry messages from one part of the body to another.

The hydra is an animal that has a sac-shaped body and tentacles that are covered in sting cells.

There are also feeding cells that release chemicals to digest the food. In more complex animals, there are even more types of cell, each specialized to carry out a specific job within the body.

CELL SHAPE AND SIZE

The shapes of cells are quite varied. Plant cells are encased in a rigid wall, so they tend to be rectangular. In contrast, animal cells have a flexible cell membrane with no cell wall, so they can be many different shapes. For example, they can be incredibly long and thin like neurones (nerve cells), doughnut-shape like red blood cells, or flat like the cells lining the inside of the mouth.

The size of cells is often related to their function. An egg cell is usually very large and is often the largest cell that an organism produces. It is packed full with food that will be used by the cell to grow once it has been fertilized. Cells involved in the uptake of materials such as oxygen or water tend to have a large surface area and they are often flat in shape. Their surface area may be further increased by the presence of microvilli, which are tiny folds and wrinkles in the membrane.

HOW BIG CAN A CELL BE?

There is a limit to how large a cell can grow. As a cell gets larger, its volume increases but the area of its surface membrane does not increase as much. This causes problems for the cell. A larger cell needs more oxygen and food but the cell membrane may not have a large enough surface area for transporting materials into the cell. Also, as the cell gets larger, the middle of the cell gets further away from the cell membrane and it takes longer for materials to be moved around the cytoplasm. This means that there is an optimum size for each particular type of cell.

NERVE CELLS

The **nervous system** is made up of cells called neurones, or nerve cells. Neurones are specialized to carry messages in the form of electrical impulses all around the nervous system. There are billions of these cells, with more than 100 billion neurones in the brain alone.

AXONS AND DENDRONS

Although neurones have a number of parts in common with all animal cells, including a cell membrane, cytoplasm, and a nucleus, their appearance is quite different from other cells. The main part of the neurone is the cell body, which contains the nucleus and most of the cytoplasm. The rest of the neurone is made up of nerve fibres. These are thin threads of cytoplasm that extend out from the cell body.

Often axons are covered with a layer of myelin – a bit like insulation around electrical wire. Myelin is made of fat, and it helps to speed up the transmission of an impulse along an axon.

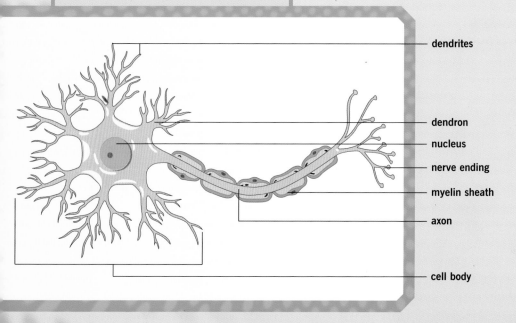

dendrites

dendron

nucleus

nerve ending

myelin sheath

axon

cell body

The longest nerve fibre is usually called the axon and the shorter ones are called dendrons and dendrites. These fibres carry the electrical messages, called impulses, along the length of the cell.

The dendrites make connections with other neurones. They pick up messages and pass them to dendrons, which carry them to the cell body. Dendrites can be located on one or both ends of the neurone. From the cell body, the message travels along the axon.

Neurones are the oldest and longest cells in the body. You have many of the same neurones for your whole life. Although other cells die and are replaced, many neurones are never replaced when they die. In fact, you have fewer neurones when you are old compared to when you are young. Neurones can be quite large – a single sensory neurone from your fingertip has an axon that extends the length of your arm.

SCIENCE PIONEERS Schleiden and Schwann

Matthias Jakob Schleiden (1804–1881) and Theodor Schwann (1810–1882) carried out some of the earliest studies of cells. In 1838, Schleiden discovered that all the various parts of plants were made of cells. His research showed that the nucleus was the control centre of the cell. Schwann recognized that some organisms were unicellular (single-celled), while others were multicellular (many-celled). He also discovered that the ovum (egg) was a single cell and he also observed that after **fertilization**, the egg divided repeatedly to develop into a complex organism.

SENSORY AND MOTOR NEURONES

There are several types of neurone in the body, two of which are the sensory neurones and motor neurones. The sensory neurones carry impulses from sensory receptors to the brain or spinal cord. Sensory receptors are cells that detect changes in the surroundings, for example a pain receptor or temperature receptor in the skin. Motor neurones carry messages from the brain or spinal cord to the muscles.

A good example of how these neurones work together is the spinal reflex. A reflex is an instant reaction to something. For example, if you touch a hot object your finger is immediately jerked away. You do not have to think about doing anything, it happens automatically. Reflexes are designed to protect you. The reflex does not involve the brain as it would take too long for the message to be carried all the way to the brain and back. Instead, reflex messages are carried along a simple loop that involves the sensory neurones, the motor neurones and the spinal cord. When you touch something hot, sensory receptors in your fingertip detect the heat and a message is sent along a sensory neurone to the spinal cord. The message is then passed on to a motor neurone. This carries the message to muscles in the arm. The muscles contract, pulling the finger away from the hot object.

THE BRAIN

The human brain is a pinkish-grey mass made up of about 10 billion nerve cells, all linked together. The brain is the control centre for movement, sleep, hunger, thirst, and virtually every other vital process essential for our survival. The largest part of the brain is the cerebrum, which is made up of two cerebral hemispheres. **Conscious** thought and memory take place here. The different parts of the cerebrum have specific functions. For example, there are areas concerned with sight, speech, memory, and personality. The cerebellum controls the co-ordination of body movements, while the medulla oblongata controls the heartbeat and breathing.

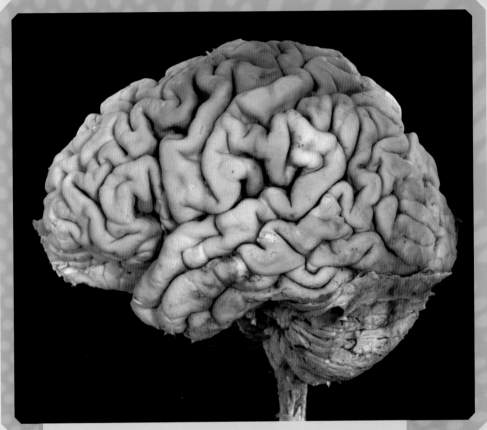

The brain of an adult human weighs about 1.5 kilograms (3.3 pounds). The surface of the cerebrum is highly folded. It is split down the middle into two halves called cerebral hemispheres. These two halves are connected to each other. The cerebellum and the medulla oblongata lie beneath the cerebrum.

Did you know..?

The brain is surrounded and protected by three membranes called the meninges. The two inner membranes enclose a fluid which acts as a shock-absorber to protect the brain from physical injury. The disease meningitis is caused by bacteria or viruses infecting the membranes. The membranes swell up and become inflamed. Some forms of meningitis are very serious and need to be treated quickly.

THE BLOOD

Blood is made up of a liquid called plasma, which contains three types of cell: red blood cells, white blood cells, and platelets. Each of these cells has a particular role in the body.

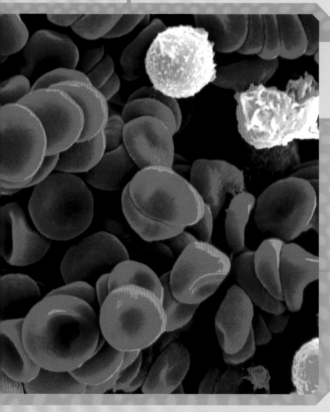

There are between 4 and 6 million red blood cells per cubic millimetre, but only between 5000 and 9000 white blood cells.

RED BLOOD CELLS

Red blood cells are the most numerous cells in the blood. Their role is to pick up oxygen in the lungs and carry it to the other cells in the body. In order to do this they contain a protein called haemoglobin which can pick up oxygen. Each molecule of haemoglobin can carry four molecules of oxygen. Haemoglobin is red in colour, which is why red blood cells are red.

The red blood cell has a round shape, but unlike most other cells it lacks a nucleus. The cell does not need a nucleus since it only has a short life span. The lack of a nucleus gives the cell a sunken appearance – a bit like a doughnut because they are thinner in the middle and thicker around the edges. The absence of the nucleus provides more space for haemoglobin, while the sunken shape provides a larger surface area for picking up oxygen. Red blood cells are very flexible, with the ability to twist and bend through the narrowest blood vessels, the capillaries. Only one red blood cell can squeeze through a capillary at a time and this means that the oxygen has time to leave the red blood cell and diffuse into the cells.

Red blood cells are made in the **bone marrow**. During the formation of a red blood cell, the cell loses its nucleus and much of its cytoplasm. The life span of a red blood cell is about 120 days, during which time it circulates continuously in the bloodstream. When the red blood cells are too old and no longer function properly, they are taken to the liver where they are broken down.

TRAINING AT HIGH ALTITUDE

When you travel high up into the mountains, the air has less oxygen and it becomes increasingly difficult to breathe. If you spend some time in the mountains, the body compensates by making more red blood cells. If there are more red blood cells, the body can pick up more oxygen. Often athletes travel to mountainous areas to train. This altitude training boosts their red blood cell count and when they return to sea level they still have the extra cells and can perform better in races.

Mountaineers climbing the world's highest mountains get breathless easily, so they have to be very careful.

Did you know..?

The blood in an adult human contains about 25 trillion red blood cells and has to replace them at the rate of about 3 million per second. Each red blood cell has about 280 million haemoglobin molecules and can pick up 1,120 million molecules of oxygen.

CELLS FOR PROTECTION AND HEALING

WHITE BLOOD CELLS

White blood cells protect the body against disease. These cells are rounded in appearance and have a nucleus. We have far fewer white blood cells than red blood cells. A drop of blood usually contains between 5,000 to 9,000 white blood cells, although this increases greatly during times of illness. They are made in the bone marrow and have a lifespan of between 13 and 20 days, after which time they are destroyed.

White blood cells are continually on the lookout for signs of disease. When germs like bacteria or viruses invade the body, the white blood cells have a number of ways they can attack. Some white blood cells produce substances called **antibodies**. The antibodies attach to germs and destroy them. Other types of white blood cells, called neutrophils, surround the germ and destroy it.

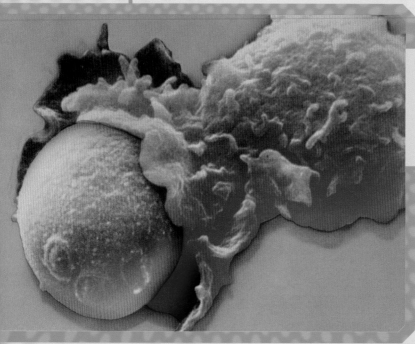

The neutrophil seeks out invading bacteria. When it finds one, its membrane flows around the bacteria and pulls it into the neutrophil cell, where it is digested.

If the germs persist and start to increase in number, the white blood cells multiply too. The presence of germs in the body causes the white blood cells to release a substance that stimulates the bone marrow into producing more white blood cells. Within a couple of hours, the number of white blood cells seeking out germs can double. Once the germs have all been destroyed, the number of white blood cells soon falls back to normal. However, a constantly high number of white blood cells may be a symptom of **leukaemia**, a **cancer** of the blood. A leukaemia patient may have as many as 50,000 white blood cells in a single drop of blood.

PLATELETS

The body has to be able to stop blood loss quickly if a blood vessel is damaged, such as when you cut yourself. Platelets play an important part in blood clotting. When bleeding starts, platelets gather at the wound and try to block the flow of blood by forming a clot. A clot begins to form when the blood is exposed to air. The platelets sense the presence of air and break apart. This releases chemicals that stimulate the formation of tiny threads which form a mesh. The mesh traps blood cells and when it hardens, it forms a scab.

SCIENCE PIONEERS blood tranfusions

Every day, blood transfusions save thousands of lives. People receive blood either during a **blood transfusion** or an operation. The first successful blood transfusion was performed by Richard Lower in 1666, when he transferred blood from one dog to another. In the following year, he carried out a successful blood transfusion on a human using blood from a sheep.

In 1795, Philip Syng Physick carried out the first known human blood transfusion in Philadelphia, in the USA.

WHY DOES A CELL BECOME SPECIALIZED?

The process of specialization is very complex. So how do cells change and become specialized for different jobs?

ANIMAL SPECIALIZATION

The first cells produced by an **embryo** are all the same. A fertilized egg cell divides into 2 identical cells, and then into 4, 8, and 16 cells, and so on. Each new cell is identical to the parent cell. The process of cell division is called **mitosis** (see page 37). However, once the embryo has about 100 cells, it enters a different stage. It starts to change shape and the cells become different and undertake new roles. As a cell becomes specialized, it loses some structures and may gain others. For example, cells that are to become red blood cells lose their nucleus, while nerve cells develop long cytoplasmic extensions. In humans, there about 200 different cell types, all of which are derived from a single fertilized egg cell.

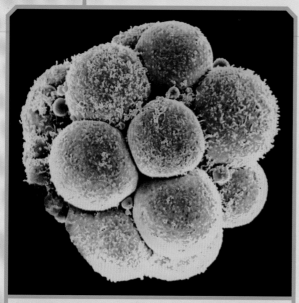

This ball of cells is an embryo. At this stage all the cells are identical, and they are dividing by mitosis to make more identical cells.

Scientists study embryos at different stages of development in order to learn about the process of change. They believe that cells know how to change from their position in the embryo and from the type of cells that surround them. For example, a cell on the outside of the embryo will become a skin cell, while a cell towards the middle changes into a cell that lines the gut. Muscle cells form from cells lying in the middle of the embryo. Chemicals may be produced that control the change too.

This photograph of a root tip shows the area where cell division is taking place. The tip of the root is protected by a root cap. Immediately behind the root cap is an area of small cells which are dividing all the time. The new cells are pushed further back and as this happens they get larger and become specialized.

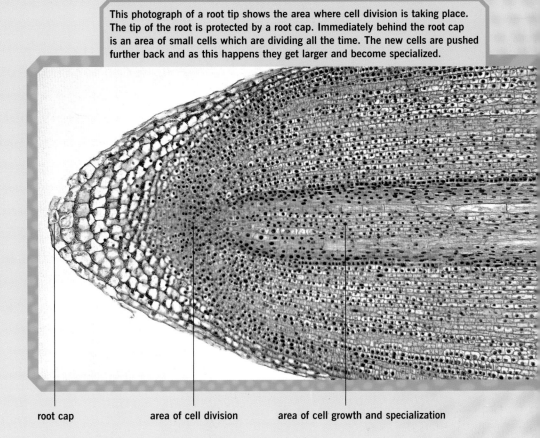

root cap   area of cell division   area of cell growth and specialization

PLANT SPECIALIZATION

Plant cells become specialized too. New cells are only made in specific areas of a plant, for example in the shoot and roots tips and in the buds. Here unspecialized cells divide repeatedly to form new cells. As new cells form, the older cells get pushed away. The cell becomes elongated in shape and a vacuole appears. As the cell gets further away, it starts to become specialized in order to carry out a particular job. For example, a sieve tube cell transports sugar. It loses its nucleus and its cytoplasm is squashed to the sides of the cell. This makes more room for moving sugar. The cell is rectangular in shape and the cell walls at the ends of the cell have tiny pores in them, allowing sugar to move from one cell to the next.

Cells, tissues, and organs

Cells in multicellular organisms are not arranged haphazardly, but organized very carefully into tissues, organs, and systems.

TISSUES, ORGANS, AND SYSTEMS

A tissue is a group of similar cells that work together to carry out a particular job, for example muscle or nerve tissue. Tissues are grouped together into organs. An organ is a structure that contains at least two different types of tissue that work together for a common purpose. There are many different organs in the body, including your liver, kidneys, and heart. Even your skin is an organ. The stomach, for example, is an organ that contains tissues that produce digestive enzymes, muscle tissue in the stomach wall, as well as connective tissue that joins all the different tissues together.

Groups of organs work together to carry out various vital functions in the body. Such a group is called a system. For example, the digestive system includes the mouth, stomach, pancreas, and small and large intestines, while the urinary system consists of the kidneys and bladder. Other important systems in the body include the blood system, reproductive system, and breathing system.

MUSCLE TISSUE

Muscles are made of tissues that work by getting shorter or contracting. This is possible because they contain fibres made up of protein. However, the process of contracting uses up a lot of energy so their cells need a constant supply of food and oxygen, which is bought in by blood vessels.

There are three kinds of muscle tissue in the human body: striated or skeletal, smooth, and cardiac muscle. Striated muscle is attached to the bones of the skeleton, like the biceps and triceps muscles in the arm. You can control these muscles. Striated muscle is named because of the way the cells are arranged, giving it a striped appearance. This tissue is unusual since it is not visibly divided into cells along its length. Instead there are bundles of long fibres. When these fibres contract, the whole muscle gets shorter.

Smooth muscle is found in a number of places in the body, including in the wall of the intestine, stomach, and the bladder. This type of muscle can contract on its own. Unlike striated muscle, you have no control over smooth muscle. The muscle in the stomach wall contracts slowly to mix the food in the stomach, while smooth muscle in the intestines contracts to push the food along.

Cardiac muscle is found only in the heart, and contracts and relaxes rhythmically throughout a person's life and never tires. It does not need nerves to make it contract. However nerves from the brain to the heart can speed up or slow down the heartbeat.

muscle fibre   striated cytoplasm   nucleus

Striated muscle is used to move the bones of the body. It consists of long fibres and is well supplied with blood vesslels which bring oxygen.

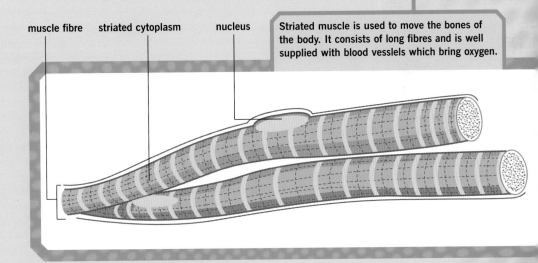

Did you know..?

Muscles make up about 40 percent of an adult's body weight.

BONE TISSUE

The human skeleton is made up of bone and **cartilage**. You may not believe it but bone is a living tissue and contains cells and blood vessels. The bulk of the bone is made up of a mineral called calcium phosphate. This mineral makes bone very hard and gives it a white appearance. There are also lots of collagen fibres, a type of protein that provides elasticity.

A bone is made up of a layer of compact bone tissue around the outside and a layer of spongy bone tissue beneath it. Spongy tissue has lots of spaces between slivers of bone and

If you examine compact bone tissue under a microscope, you find that it is made up of rings of cells. The cells are embedded in a substance called the matrix and this is made up of calcium phosphate and collagen that give the bone its strength.

these are filled with blood. The blood brings in oxygen and food for the cells in the bone tissue. Also the spaces help to make the bone lighter. In the centre is a space that is filled with bone marrow. Bone marrow is a soft tissue that has a lot of blood vessels. It is here that the red blood cells, white blood cells, and platelets are made.

LUNG TISSUE

The role of lung tissue is to assist the uptake of oxygen by red blood cells. The most important elements of the lung are tiny sacs called alveoli. The wall of the alveoli is made up of a single layer of epithelial cells. These cells are flat with a large surface area. Gases can pass easily across them. Wrapped around the outside of the alveoli are lots of tiny blood vessels called capillaries. Air is drawn into the lungs, filling all the alveoli. Oxygen in the air diffuses through the walls of the alveoli into the blood capillaries. Some of alveolar cells secrete a liquid that helps oxygen to diffuse more easily across the cells. Carbon dioxide in the blood diffuses in the opposite direction. It leaves the blood and diffuses through the wall of the alveoli into the air which is breathed out.

PREMATURE BABIES

One of the most common problems facing a premature baby is difficulty breathing. Because it has been born too early, the baby's lungs have not developed properly and they do not produce enough of an important substance called surfactant. Surfactant allows the inner surface of the lungs to expand properly when the baby starts to breathe air for the first time. Nowadays, women about to give birth to a premature baby are given medication just before delivery to help the baby breathe. Immediately after birth and several times later, the baby is treated with artificial surfactant.

SKIN – THE LARGEST ORGAN

The skin is the largest organ in the human body. Its roles are to protect the body, provide a barrier against micro-organisms such as bacteria, and to help control body temperature.

The skin is composed of three layers: the epidermis, dermis, and subcutaneous layer. The epidermis is the top layer of skin. It consists of cells tightly packed together, providing a barrier between the inside of the body and the outside world. This is called epithelial tissue. There are many epithelial layers and new cells are constantly formed at the bottom, pushing the cells upwards. As this happens, the cells become fat and hard and eventually die, forming a dead layer at the surface.

The skin is a complex organ which is specialized to perform several key functions.

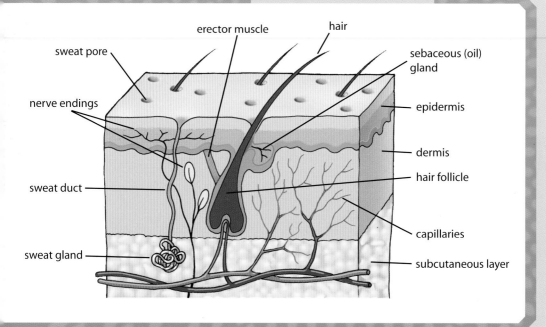

Below the epidermis lies the dermis. The dermis is made up of tough connective fibres of collagen and elastin that support the skin and give it elasticity. As a person gets older, these fibres lose their elasticity and the skin becomes loose and wrinkled.

The subcutaneous layer is beneath the dermis and consists mainly of cells that are packed full of large drops of oil. This is called adipose tissue, more commonly known as fat. Adipose tissue helps cushion the skin and provide insulation against the cold.

CASE STUDY treating burns

Burns can be horrific injuries. Some burns just damage the epidermal tissue on the surface of the skin. The skin can repair these burns. But the most severe kind of burn is one that damages the dermis that lies under the epidermis. The skin cannot recover from these types of burns and so doctors have to perform skin grafts. This is a process that involves taking healthy skin from one part of the body and attaching it to the skin that has been injured. If a person has suffered burns over much of their body, they may not have enough healthy skin left to supply a graft. Now doctors can use an artificial skin that consists mostly of collagen fibres. The damaged skin is removed and the artificial skin is laid over the wound. The artificial skin helps the skin cells to grow new fibres. In time, new blood vessels grow into the area and a new skin forms over the damaged area.

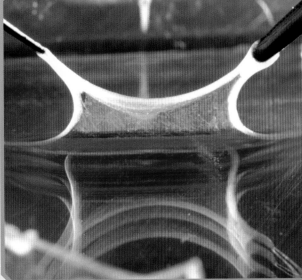

This is artificial epidermis that will be used for a skin graft. This cutting-edge technology helps people with serious injuries, reducing the long-term physical effects of burns.

PLANT TISSUES

Just as in animals, plant tissues are groups of similar cells that carry out the same role. There are several different tissues in the leaf, including the epidermal and mesophyll tissue. There are tissues involved with the transportation of materials around the plant, as well as different tissues in the flower and root.

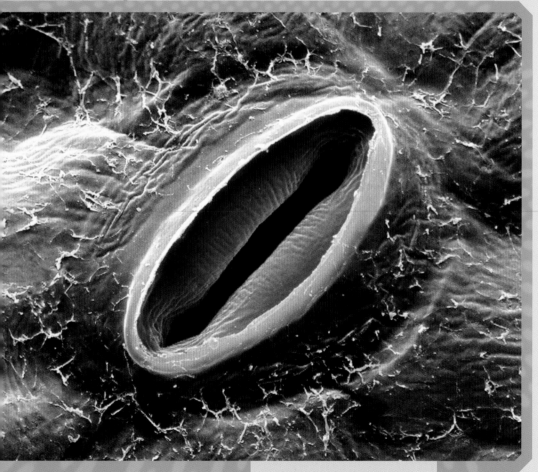

Most leaves have stomata in their lower epidermis. Each stoma is a gap in the epidermis and it is surrounded by two sausage-shaped guard cells.

EPIDERMAL TISSUE

A plant's epidermal tissue forms the outermost layer and is the plant equivalent to animal skin. These cells are flat and form a protective layer. Often, they have a layer of cuticle on their outermost surface. This is a thin, waxy layer that waterproofs the leaf surface and helps to reduce the amount of water that evaporates from the leaf surface. In most leaves, there are gaps in the epidermis, especially the lower epidermis, called stomata, between the epidermal cells. Each stoma is surrounded by two specialized cells called guard cells, which can open and close the stoma. The stomata allow gases such as carbon dioxide and oxygen to enter and leave, and water vapour to escape.

PACKING TISSUE

Simple packing tissue fills the spaces in the plant. The cells are rounded and very firm, a bit like balloons packed together with tiny air spaces in between. They provide support and a place to store materials. For example, packing tissue in the root is filled with starch.

VASCULAR TISSUE

Running through the plant are the vascular tissues made up of xylem and phloem. These tissues are important as they transport water and food around the plant. They also provide essential support, especially the xylem. Xylem contains cells called xylem vessels that are hollow tubes. The walls of these cells are made of lignin, a very strong material. Lignin makes the cell wall impermeable so nothing can pass through. Once a cell wall is completely filled with lignin, the contents of the cell die. This creates the hollow tube that forms the xylem vessels. Water move through these tubes. Hollow tubes are surprisingly strong and so the xylem helps to hold the plant's stem erect. Phloem contains sieve tube cells that are adapted to carrying food materials around the plant. These cells have pores in their end walls so that food can pass from one cell to another. Sieve tubes have lost their nucleus so there is more space for moving food.

LEAVES, STEMS, AND ROOTS

The main organs of the plant are the root, stem, and leaf.
A plant organ is made up of several different tissues working
together.

LEAVES

The leaf is very important as it contains most of the
photosynthetic tissue that makes the food for the plant.
Leaves contain a network of veins, which are bundles of tubes
called xylem and phloem. Xylem carries water and minerals
into the leaf cells, and phloem carries sugars out to the plant.
A plant could not survive without its leaves. The leaf has a
large surface area and is usually flat. This means that gases
such as carbon dioxide do not have to travel very far to pass
from the air to the cells.

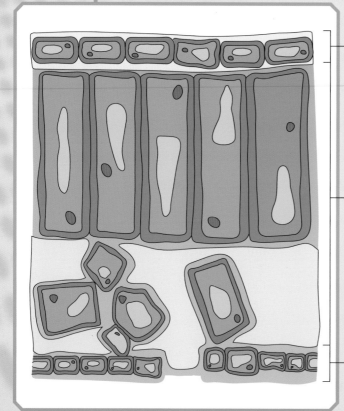

Upper epidermis – single
layer of cells, often with
a thin waxy layer called
a cuticle over the upper
surface to reduce water loss.

Mesophyll – this is made
up of two layers; the
palisade mesophyll above
and the spongy mesophyll
below. The palisade
mesophyll cells are brick-like
in shape and are usually
arranged in two or three
rows below the upper
epidermis. The spongy
mesophyll is made up of
rounded cells, loosely
packed together with lots
of large air spaces.

Lower epidermis – single
layer of cells with pores
called stomata. Each stoma
consists of a pair of guard
cells surrounding an opening.

STEMS

The stems support the leaves of the plant. The vascular bundles of xylem and phloem are arranged in a ring in the cortex to provide more support.

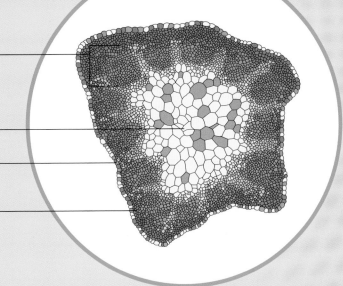

Vascular bundle – these give support to the plant, and transport materials around the plant in xylem and phloem tubes.

Pith – the cells at the centre of the stem.

Epidermis – this is a single layer of cells that protects the stem.

Cortex – the space between the epidermis and the vascular bundles is filled with simple rounded cells that act as packing tissue.

ROOTS

Roots are organs that perform two main jobs for the plant. Firstly, they support the plant by anchoring it in the ground. Vascular bundles of xylem and phloem tubes are located in the centre of the root to give more support. Roots also take up water and minerals, through specially shaped root hair cells. These are elongated epidermal cells. The water is taken up by the tiny root hairs and transported around the plant in the xylem tissue. The root hairs are positioned behind the root tip.

Making new cells

Growth is one of the characteristics of life. Growth involves an increase in size, either through an increase in the number of cells or through an increase in the size of each individual cell.

WHY DO CELLS DIVIDE?

There is a limit to how large a cell can become, so in order to grow larger, an organism needs to make more cells. The new cells are produced by a cell dividing to form two cells. This process is called mitosis. In this type of cell division, the two new cells have the same number of chromosomes as the parent cell and are identical to it and to each other.

GROWTH

In multicellular animals, growth takes place all over the animal's body until the animal reaches full size. In complex animals such as mammals, a few cells retain the ability to grow and divide. For example, cells in the skin continue to divide to provide a constant supply of new cells as old ones are lost. Cells in the bone marrow also continue to divide to produce new blood cells. All the other cell types lose their ability to divide. This means that some cell types cannot be replaced. An example of this is when brain cells die they are not replaced.

In plants, growth occurs in specialized areas called meristems. Cell division in these regions increases the length of the shoot and root. There is also a small area of growth located between the xylem and phloem in the vascular bundles. Cell division here makes the stem wider.

MITOSIS

In this type of cell division, one cell divides to form two identical cells. First the nucleus divides into two and this is followed by the rest of the cell dividing into two. As the nucleus starts to divide, the chromosomes become visible as threads within the nucleus. The DNA has been copied, so each chromosome consists of two identical threads, joined together at one point. The chromosomes move to the middle of the cell where they line up. Then they move apart, each thread going in the opposite direction. Finally the cytoplasm divides and the two new cells are formed. A nucleus forms in each of the new cells.

At the start of mitosis the nuclear membrane disappears (step 1) and the spindle appears (step 2). The chromosomes move to the middle of the cell (step 3) and the two threads of each chromosome are pulled apart (step 4). These move to opposite ends of the cell. Finally, the nuclear membrane reforms (step 5), and there are then two identical cells.

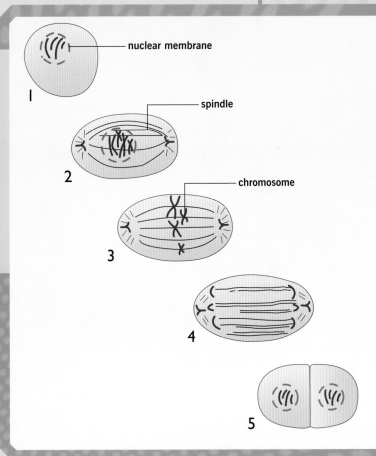

nuclear membrane

spindle

chromosome

REPRODUCTION

Organisms reproduce in order to produce new individuals. There are two forms of reproduction, sexual and asexual reproduction.

SEXUAL REPRODUCTION

Mammals reproduce using sexual reproduction, which involves two individuals, a male and a female. The male produces sex cells called spermatozoa that fertilize the egg cells produced by the female. The new individual is different from both of its parents.

ASEXUAL REPRODUCTION

Asexual reproduction involves a single parent and is common in single-celled organisms such as bacteria and amoebas. The parent organism reproduces by simply dividing into two. This is called binary fission. Both new cells are identical to the parent. The two new cells feed and grow, and then, if conditions are favourable, they will divide again.

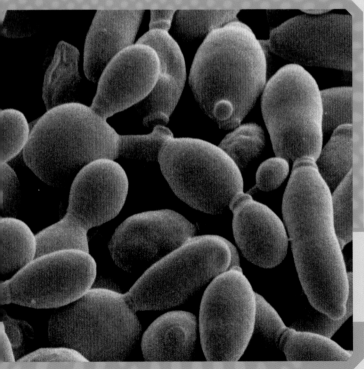

A new yeast cell forms as a bud on the side of a fully grown cell. Once is it large enough, it breaks away to become an independent cell.

An amoeba may divide only once a day, whereas a bacterium may divide as much as once every 20 minutes, given the right conditions. One bacterium could become one million in just seven hours. Yeast and hydra reproduce asexually by budding. The bud grows and then separates to become an individual.

Many plants also reproduce asexually, and gardeners make use of this when they take cuttings. A cutting is simply a length of shoot removed from a plant, placed in soil, and allowed to grow roots. It grows into a new plant which is identical to the parent plant.

SCIENCE PIONEERS Walther Flemming

Zoologist Walther Flemming developed a staining technique that enabled him to watch what happened to chromosomes during cell division. The new staining techniques made it possible for Flemming to follow the process of cell division, which he named mitosis. In 1882, his results were published in the book *Cytoplasm, Nucleus and Cell Division*. It was another 20 years before the significance of Flemming's work was truly acknowledged with the rediscovery of Gregor Mendel's rules of **heredity**.

Did you know..?

The largest and smallest cells in the human body are the **gametes**, or the sex cells. The female sex cell, the egg, is about 35 μm in diameter, which is just visible to somebody with really good eyes. The male sex cell, the spermatozoon, is only about 3 μm in diameter, and is the smallest cell of the human body.

CLONING

A **clone** is an individual that is genetically identical to another individual. This means that every single bit of DNA is exactly the same. Cloning is often thought of as something that only happens in a laboratory, but it can occur naturally. Identical twins are clones because they are formed when a newly fertilized egg splits into two, and each part develops into a new individual. Since the twins are formed from the same cell, they are genetically identical.

Individuals that have been produced by asexual reproduction are clones too. This includes plants that reproduce asexually. Strawberry plants produce shoots called runners that grow along the ground. New plants form at intervals along this shoot and they are all clones of each other and of their parent plant.

IN THE LABORATORY

Recently, the term cloning has been associated with making copies of animals in the laboratory. The best known clone was Dolly the sheep. Normally, sheep reproduce sexually so the offspring are different from their parents. Dolly was different – she was an identical copy of her mother.

There are two methods of laboratory cloning. One method is called "embryo cloning". It is based on the natural process by which identical twins are formed. A fertilized egg is taken and, once the embryo has started to grow, the cells are pulled apart and allowed to develop into individual embryos. The offspring are clones of each other, but are different from their parents.

The other method is called nuclear transfer and is how Dolly the sheep was created. In the first stage, scientists take an unfertilized egg and remove its nucleus. Then a cell is taken from the animal that is to be cloned. This is called the donor cell. The nucleus is taken out of the donor cell and placed inside the empty egg. It is given a tiny burst of electricity to get it to start dividing and form a ball of cells. Then the new embryo is implanted in the **uterus** of a female animal, where it develops until it is ready to be born.

KEY EXPERIMENT creating Dolly

Dolly the Sheep was created by a team led by Ian Wilmut at the Roslin Institute in Scotland in 1996. Dolly's mother was a six-year-old Finn Dorset ewe (female sheep). The process was incredibly difficult and the team experienced many failures. They tried replacing the nucleus in the egg 277 times and they only obtained 29 embryos. All 29 embryos were placed in **surrogate** ewes but only one survived – Dolly.

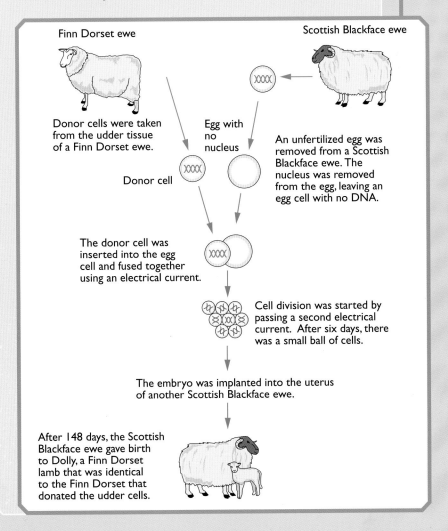

Finn Dorset ewe

Scottish Blackface ewe

Donor cells were taken from the udder tissue of a Finn Dorset ewe.

Egg with no nucleus

Donor cell

An unfertilized egg was removed from a Scottish Blackface ewe. The nucleus was removed from the egg, leaving an egg cell with no DNA.

The donor cell was inserted into the egg cell and fused together using an electrical current.

Cell division was started by passing a second electrical current. After six days, there was a small ball of cells.

The embryo was implanted into the uterus of another Scottish Blackface ewe.

After 148 days, the Scottish Blackface ewe gave birth to Dolly, a Finn Dorset lamb that was identical to the Finn Dorset that donated the udder cells.

STEM CELLS

Stem cells are special cells that retain the ability to divide, even when most other cells no longer can. When they divide, one cell remains a stem cell but the other cell develops into a different type of cell. A stem cell in a human embryo can divide and form any one of more than 200 different types of cells. As the individual gets older, the ability to make cells decreases and adult stem cells can only make a few types of cell.

MAKING AND REPLACING

The bone marrow is particularly rich in stem cells. It is here that stem cells make red blood cells, some types of white blood cell, and platelets. There are also stem cells in other organs such as the skin and the intestines. Some tissues have stem cells that lie in wait until they are needed. If the tissue is damaged, the stem cells become active and repair the damage, for example in muscle tissue. However, brain tissue does not have any stem cells, so it cannot make new cells even when some die.

In the future, it may be possible to cure many diseases by growing a supply of new cells from stem cells in the laboratory. This may help people who have suffered from extensive burns, diabetes, liver disease, heart disease, and various brain disorders such as Parkinson's Disease. The treatment is called cell replacement therapy. In this kind of therapy, the damaged organ or tissue is injected with cells that have been grown in the laboratory from stem cells.

STEM CELL RESEARCH AND ETHICS

Stem cells taken from embryos are the most useful to medical research as they can be used to make a full range of cell types. They are taken from human embryos no more than 14 days old. At this point the embryo is just the size of a full stop and contains about 100 cells. The cells have not yet started to become specialized. A few stem cells are removed and grown separately to make lots more stem cells.

The use of stem cells could offer major medical breakthroughs, but it raises many **ethical** issues that governments, scientists, and the public continue to debate. Some people believe that the use of embryonic stem cells is wrong because an embryo is a potential human life. Others believe that stem cell research, if used properly, can offer an ethical and beneficial new path in medical research. The rules governing experimentation on human embryos are very strict. Many countries have gone further and banned human stem cell research completely. However, it may be possible to avoid all these problems by taking adult stem cells from the bone marrow.

The round cells in this picture are embryonic stem cells. Research that involves the use of human stem cells is controversial because it involves the destruction of a human embryo.

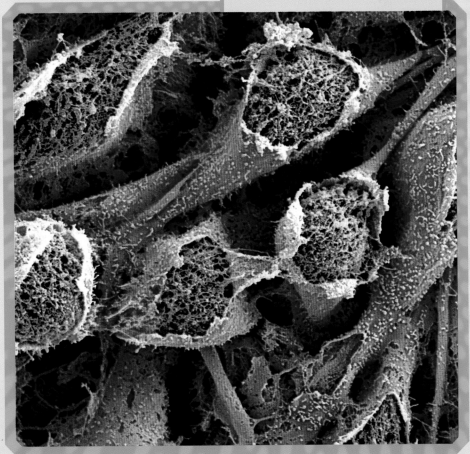

CELLS OUT OF CONTROL

Every minute, ten million cells divide in the human body. Normally, cell division takes place in an orderly way. But sometimes a cell goes out of control and starts to divide over and over again, creating a mass of cells called a tumour. This can sometimes lead to cancer.

CANCER

Cancer develops from a single cell that has mutated and the genetic material which carries the body's hereditary instructions has been changed. Instead of maturing normally and then dying, cancerous cells reproduce without control. Some divide very quickly while others divide more slowly but they all share one feature – they never stop dividing. In fact, if a cancerous cell is removed from the body and grown in the laboratory with plenty of nutrients it would continue to grow and divide forever.

Scientists are trying to find out what causes a cell to start dividing in an uncontrolled way. In many cases, it has been linked to chemicals in the environment called carcinogens that cause the cell to stop working normally. For example, the chemicals in cigarette smoke such as tar and nicotine may cause lung cancer by interfering with the **genes** that control normal growth in the cells of the lungs.

This picture, taken with a scanning electron microscope, shows a single breast cancer cell.

CANCER IN THE GENES

Sometimes cancers can be caused by genetic factors that are inherited. For example, some breast cancers run in families and the women in such families inherit certain genes from their mother that make them more likely to get breast cancer.

RECENT DEVELOPMENT bone marrow transplants

Leukaemia is the name given to a group of cancers that affect the blood. There are different forms of leukaemia and they all involve the production of abnormal white blood cells. Most cancers, including leukaemia, are usually treated with **chemotherapy** and **radiation therapy**. These are treatments that target cells that divide rapidly. However, healthy stem cells in bone marrow can be damaged by these treatments, especially high dose chemotherapy. Without healthy bone marrow, the patient cannot make new blood cells, so patients are given a bone marrow transplant (BMT) to replace the stem cells. To minimize potential side effects, doctors use bone marrow that matches the patient's own bone marrow as closely as possible. Once the patient has had their chemotherapy or radiation treatment, they receive the bone marrow from the donor.

Did you know..?

The word cancer was first used by the ancient Greek doctor Hippocrates. He thought the way tumours attached to parts of the body reminded him of crab claws so he named the disease *karkinos*, the Greek word for crab. This word translates as carcinoma in English, meaning a cancerous growth.

MEIOSIS – A DIFFERENT TYPE OF DIVISION

Egg cells and spermatozoa are gametes and they are different from all other cells in the human body. Most body cells contain 46 chromosomes but the gametes have just 23 chromosomes. Gametes are produced from cells that contain 46 chromosomes, but the number of chromosomes in the new gamete is halved in a type of division called meiosis. Meiosis is similar to mitosis, but it involves two divisions.

During the first division, the nucleus disappears and the chromosomes become visible. The 46 chromosomes organize themselves so that they are arranged in pairs. The pairs are separated and they move to opposite ends of the cell. Now the rest of the cell divides. The two new cells each have 23 chromosomes. The next division starts immediately and is identical to what occurs during mitosis (see page 37). The chromosomes gather in the middle of the cell. Each of the 23 chromosomes consists of two threads, which separate and move to opposite ends of the cell. In the final step, the cell divides. By the end of meiosis, there are four cells, each of which contains 23 chromosomes.

When an egg cell with 23 chromosomes is fertilized by a spermatozoon that also has 23 chromosomes, the new embryo ends up with 46 chromosomes. If the gametes were not produced in this way, the number of chromosomes would keep on increasing.

PREGNANCY AND OLDER MOTHERS

Mistakes can happen during meiosis, causing the resulting gametes to have chromosomes that are missing bits, or there may be extra chromosomes. Older women who become pregnant are more likely to have a baby that has a chromosome defect. This is because of the age of the egg cell. The egg cells are formed in the ovary of a woman while she is still a **foetus** in her mother's uterus. Many years later, some of these egg cells may be fertilized. Due to the long period of time between the formation of the egg cell and its fertilization, there are more chances that something may go wrong.

Some babies are born with Down's syndrome. They have 47 chromosomes rather than the normal 46. A baby with Down's syndrome has a different appearance and may have learning difficulties. An older mother is often offered a test for abnormalities like Down's syndrome when she is about 15–20 weeks pregnant. This is called an amniocentesis test. A needle is inserted through the wall of the uterus and cells from the baby are extracted. These cells are examined under a microscope for abnormalities.

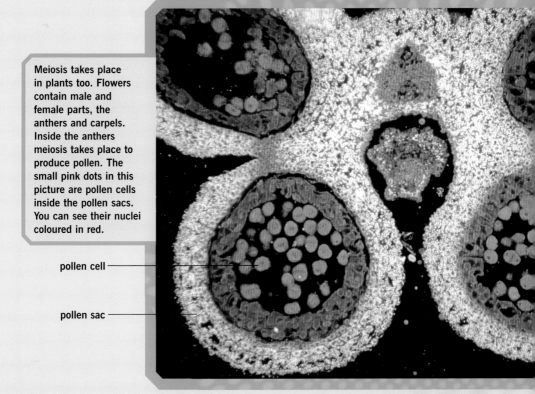

Meiosis takes place in plants too. Flowers contain male and female parts, the anthers and carpels. Inside the anthers meiosis takes place to produce pollen. The small pink dots in this picture are pollen cells inside the pollen sacs. You can see their nuclei coloured in red.

pollen cell

pollen sac

Bacteria

Bacteria are tiny single-celled organisms that are found everywhere, from hot pools of mud to the polar ice caps. There are lots of them too – several million are probably living in your mouth at this very moment. They feed on a vast range of materials, from sugar and starch to sulphur and iron. Some are extremely tough, like the bacteria called *Deinococcus radiodurans* which can withstand blasts of radiation 3000 times greater than would kill a human being.

BACTERIAL CELLS

A bacterial cell is much simpler than a human cell. It consists of just an outer membrane and the cytoplasm. It doesn't even have a nucleus. At the centre of the cell is a ball of DNA. If this DNA was stretched out into a single long strand, it would be about 1,000 times longer than the bacterial cell itself.

Often there is a capsule around the outer membrane of a bacterium to give additional protection and prevent it from drying out.

During the 1960s, bacteria were found in the hot springs of Yellowstone National Park. These bacteria can survive very high temperatures and have been studied in the laboratory for many years.

Some bacteria have long thread-like tails, called flagella, attached to the outside of the cell. They use them to move around. Bacteria can reproduce by binary fission. If provided with plenty of food and space they can divide once every 20 minutes.

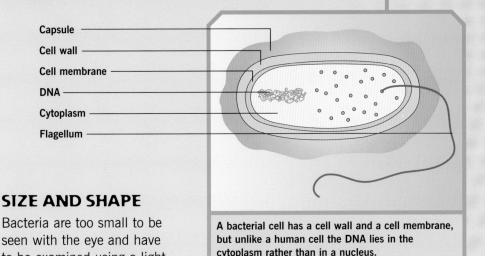

Capsule

Cell wall

Cell membrane

DNA

Cytoplasm

Flagellum

SIZE AND SHAPE

Bacteria are too small to be seen with the eye and have to be examined using a light microscope. For example, the

A bacterial cell has a cell wall and a cell membrane, but unlike a human cell the DNA lies in the cytoplasm rather than in a nucleus.

bacterium *Escherichia coli* is about 1–2 micrometres long, which is about one-hundredth the size of a human cell.

Bacteria have distinctive shapes, and this is one of the ways by which they are identified. Spherical bacteria are called cocci. They can occur singularly, in pairs, or in chains. For example, *Streptococcus*, the bacterium that causes sore throats, forms long chains. Rod-shaped bacteria are called bacilli, and they include *E. coli* and *Salmonella*. The bacterium that causes cholera, *Vibro cholerae*, has a shape that resembles a bent rod.

SURVIVAL

Many bacteria can survive extreme environmental conditions such as drought or freezing temperatures. They do this by forming a protective coat around themselves and becoming spores. They remain as spores until the conditions become more favourable. Then the coat splits and the bacterium can become active again. Some bacteria can survive 50 years or more in this condition.

GENE TECHNOLOGY

Bacteria, yeast, and moulds make useful products, including proteins, food such as yoghurt, and chemicals such as alcohol. To obtain these products, the bacteria or yeast are grown in large quantities in vats and the product extracted and purified. One important product is the **antibiotic**, penicillin, which is made from the mould (type of fungus) called *Penicillium*.

MAKING LIFE EASIER

With new developments in gene technology, it is now possible to change the DNA of organisms such as bacteria so that they can make something that they would not normally produce. One such product is insulin. Insulin controls the level of glucose (a sugar) in the blood. However, people who suffer from diabetes are unable to produce enough insulin and they cannot control their blood glucose levels. If untreated, this can cause serious medical problems, such as heart disease, kidney damage, and blindness. People with diabetes have to be given injections of insulin to help them control their glucose levels. In the past, the insulin came from other mammals, and this caused side affects. Now bacteria have been altered so that they can make human insulin. This doesn't cause so many side effects and has helped many sufferers of diabetes.

One product of a genetically modified yeast is used in cheese making. Traditionally, cheese is made by mixing milk with a substance called rennet that causes the milk to separate into solids called curds, and a liquid called whey. The curds are removed, pressed into shape, and become cheese. Rennet contains enzymes that are obtained from the stomach of calves after they have been slaughtered. Many vegetarians do not like the idea of eating cheese that contains this animal product. Now there is an alternative to rennet, called chymosin that is obtained from genetically altered yeast cells.

Gene technology is incredibly useful. There are many other examples of how gene technology can help us, including new **vaccines** for the disease hepatitis, which damages the liver. They can also be used in the creation of human growth **hormones**. Human growth hormones are given to children who are not growing properly and are shorter than normal. Previously this hormone could only be obtained from the bodies of people who had died and it was incredibly expensive, so very few people could be treated. Now bacteria have been altered to make this hormone.

This is a fermentation unit (bioreactor) which uses genetically-modified bacteria to produce proteins for medicine.

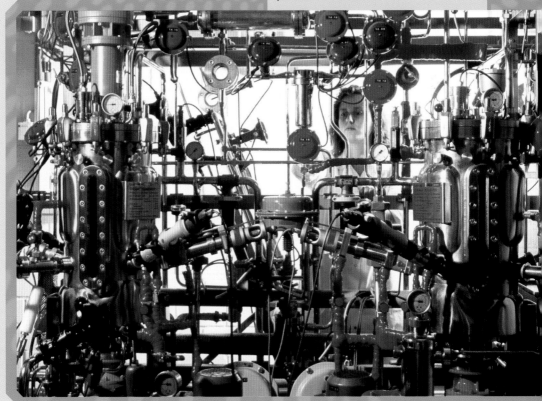

Did you know..?

Bacteria live on or in just about every material and environment on Earth. Each square centimetre of your skin has on average 100,000 bacteria living on it. A single teaspoon of topsoil contains more than a billion bacteria.

BACTERIA AND DISEASE

Bacteria are responsible for many of the diseases that affect people. They get into the body through the mouth, nose, or cuts to the skin. Once inside the body, they attack the cells. Some stick to the outside of cells and damage them. Other bacteria produce toxins or poisons that harm the body. For example, *Salmonella* bacteria that cause food poisoning release a toxin that passes into the blood stream and causes diarrhoea and fever.

Once bacteria are in the body, they start to multiply very quickly. The time between infection and showing signs of the disease is called the incubation period. This may be a few hours, several days, or even weeks, depending on the type of bacteria.

It is easy for bacteria to spread when an infected person sneezes in a crowded place.

SPREADING DISEASE

Many bacterial diseases are contagious, which means the bacteria spread from person to person. Bacteria can be spread through droplets in the air. When an infected person sneezes, millions of bacteria are shot into the air and may be inhaled by another person. Colds and flu are spread in this way. Bacteria may infect drinking water or food. Cholera is a major health threat after natural disasters such as earthquakes or flooding because sewage gets washed into rivers and this water is then used for drinking, bathing, and cleaning food utensils. Animals may carry bacteria, for example the bacteria that cause plague are carried by rat fleas.

Each year, millions of people die from bacterial diseases. The biggest killer is tuberculosis (TB), a disease that affects the lungs. It kills as many as 3 million people every year. A major cause of death of young children in the developing world is diarrhoea, a disease caused by a range of organisms, including bacteria. The bacteria are spread in contaminated water. Diarrhoea results in the body losing a lot of water and the child becoming dehydrated. It is treated using rehydration salts.

SCIENCE PIONEERS Robert Koch

One of the most important bacteriologists at the end of the 19th century was the German, Robert Koch. He spent a lot of time studying bacteria under a microscope. Koch established that a number of diseases were caused by bacteria. He took blood samples from people who had a particular disease, and allowed bacteria from the blood to grow. Then he injected the bacteria into mice and found that the mice got the same disease. In 1905, he received the Nobel Prize in Physiology or Medicine for his work in developing a test for tuberculosis.

ANTIBIOTICS

One way to fight a bacterial disease is to use antibiotics, which are chemicals that kill bacteria. The very first antibiotic was penicillin. Now there are many types of antibiotics, which work in different ways. Some attack the bacterial cell wall, making it weaker and allowing the contents of the cell to leak out. Some act by slowing down bacterial growth. Others cause the bacteria to clump together, so that they are more easily destroyed by white blood cells.

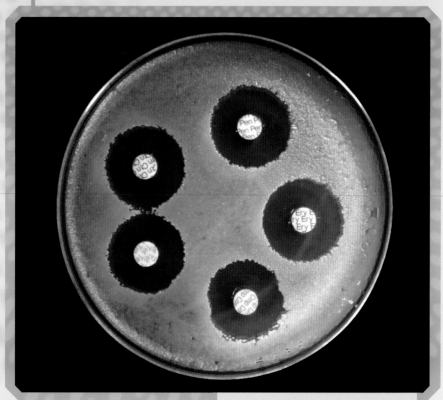

Bacteria are growing on the surface of the nutrient agar jelly. The discs have been soaked in an antibiotic which spreads out from the disc. The antibiotic has stopped the growth of bacteria, as seen by the clear area around the discs.

PROTECTING AGAINST BACTERIA

The body protects itself naturally by building up immunity against a disease. If you suffer from an illness, white blood cells increase in number. Some of the white blood cells seek out the bacteria and kill them. Other types of white blood cell produce chemicals called antibodies. Antibodies attach themselves to invading bacteria and destroy them. There are a lot of different antibodies, and each one works against a different type of bacteria. Sometimes it takes a while to make enough of the right antibodies and the bacteria have a chance to breed and increase in number. This causes the symptoms of the disease to appear. Eventually the antibodies overcome the bacteria and the person recovers. However, the next time that type of bacteria invades, the body recognizes it immediately. Lots of antibodies are made quickly so the bacteria are destroyed before they can cause the disease. The person has become immune to the disease.

Nowadays, children are **vaccinated** against many diseases. A vaccine contains either a dead or weakened version of the bacterium or virus. This is injected and the body reacts by producing lots of antibodies. You don't get the disease because the organisms have been weakened or killed. But the next time this organism invades your body, antibodies are produced quickly and the organism is killed. Sometimes a single vaccination is sufficient to protect you for life, but often a second vaccination is needed. Booster vaccinations may be given every few years as protection wears off.

SCIENCE PIONEERS Alexander Fleming

Alexander Fleming was the first to discover substances that could kill bacteria. In 1928, he was growing bacteria in the laboratory and just by chance he noticed that a mould had killed some of the bacteria. He identified the mould as *Penicillium*. It took 12 years before the substance produced by the *Penicillium* could be extracted and used medically. The new miracle drug was called penicillin.

SUPERBUGS

Bacteria reproduce quickly. If a change occurs in their DNA, the change is soon passed on to new generations of bacteria. Some of these changes make the bacteria more **resistant** to drugs such as antibiotics. Antibiotics are the main weapon against bacteria but they have been overused and now many types of bacteria are resistant to them.

A big problems in hospitals at the moment is the rise of the so-called superbug, MRSA – methicillin resistant *Staphylococcus aureus*. This bacterium has become resistant to commonly used antibiotics. These bacteria are commonly found on the skin and in the throat and do not cause too many problems for healthy people. However, they can cause severe infections in sick or weak patients in hospital as the bacteria can enter their body through a burn, wound, or surgical scar. Antibiotics can be used to combat the infection, but powerful doses are required and fewer and fewer antibiotics are effective. The best way to fight these bacteria is good hygiene, for example washing hands before touching a patient and making sure the hospital is clean.

The spread of bacteria in hospitals can be reduced by the use of clean protective clothing, such as gowns and face masks, and the cleaning of hands with antibacterial soap.

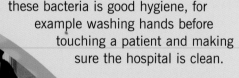

CASE STUDY viruses

Most people get a cold or flu at some point in their life. These diseases, and many others, are caused by viruses. Viruses are extremely small, usually a thousand times smaller than a bacterium – just 17 to 300 nm in length. They can only be observed using an electron microscope. A virus particle consists of a protein coat wrapped around a strand of DNA or **RNA**.

Viruses do not have a cell membrane or even a nucleus or cytoplasm. They are nothing like other organisms. In fact, most of the time, they exist as inactive particles that are hardly living. Viruses need to reproduce, but they do not have the machinery to build new particles themselves, so they invade a cell and instruct it how to make new viral particles. The cell that they invade is called the host and each type of virus has its own preferred type of cell to infect.

This is a virus viewed through an electron microscope.

The viral particle attaches to the outside of its host cell and injects its DNA or RNA into the cell. Once inside, the viral DNA or RNA takes over the operation of the cell, directing the cell to make more viral particles. Eventually the cell is so full of new viral particles that it bursts open, freeing the particles to attack new cells. Using this system, the virus can reproduce and infect other cells at an amazing rate. Antibiotics have no effect on a virus because a virus is not alive. There is nothing to kill! The best way to fight viral diseases is **immunization**.

Future cells

In recent years, scientists have been able to discover more about the structure of the cell using the latest electron microscopes. In the future, it is likely that most of the new research will focus on how cells can be controlled or altered. For example, how stem cells can be used to treat human disease and how bacterial cells can be altered to make new products.

NEW SOURCES OF STEM CELLS

Ethical concerns about the use of human embryos in stem cell research continue to be debated. However, some scientists are avoiding this problem by taking stem cells from the umbilical cord that links a baby to its mother's placenta. A new source of stem cells is important as more and more uses are being proposed for them.

Other researchers have found a way of treating bone marrow stem cells so that they are more likely to develop into brain cells once they are injected into the brain. This may lead to treatments of diseases such as Alzheimer's disease.

Another possible alternative to using embryonic stem cells may be a method that tricks human egg cells into dividing even though they have not been fertilized. These "embryos" contain just the chromosomes from the mother and so cannot develop into babies. The tricked eggs are allowed to divide for four or five days until they reach 50 to 100 cells. However, research is at the early stage and no-one is sure that stem cells produced in this way will behave normally.

MAKING NEW ORGANS

Researchers are working on finding ways of making new organs for transplant operations in the laboratory, avoiding the need to take organs from people who have just died. One of the first organs made in the laboratory was a dog's bladder. The bladder was transplanted into a dog where it worked for the next eleven months. It may be possible to use similar methods to make human organs in the future.

CLEANING UP THE ENVIRONMENT

Around the world there are areas of industrial land with soil contaminated by toxic chemicals. The land cannot be re-used until the soil is treated. One way to solve the problem is to remove the soil but that would create a large quantity of waste that would need to be burned or buried elsewhere. A better way would be to treat the soil where it is and this may be possible using bacteria that have been modified just for this purpose.

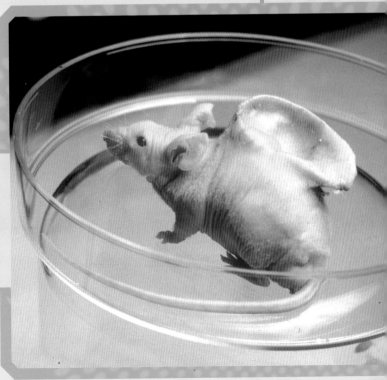

This human ear was grown on the back of a mouse in the laboratory. A plastic scaffold was inserted together with human cartilage cells. Once formed, the ear was removed from the mouse without killing it.

Further resources

MORE BOOKS TO READ

Greenberg, Keith Elliot, *Stem Cells* (Blackbirch Press, 2003)

Harper, Janet, *Cells* (Letts Educational, 2005)

Stockley, Corinne, *The Usborne Illustrated Dictionary of Biology* (Usborne Publishing, 2005)

Nature Encyclopedia (Dorling Kindersley, 1998)

USING THE INTERNET

Explore the Internet to find out more about cells and cell function. You can use a search engine, such as www.yahooligans.com or www.google.com, and type in keywords such as *cell structure*, *electron microscope*, *osmosis*, *DNA*, or *mitosis*.

These search tips will help you find useful websites more quickly:

- Know exactly what you want to find out about first.
- Use only a few important keywords in a search, putting the most relevant words first.
- Be precise. Only use names of people, places, or things.

Disclaimer

All the internet addresses (URLs) given in this book were valid at the time of going to press. However, due to the dynamic nature of the Internet, some addresses may have changed, or sites may have ceased to exist since publication. While the author and publishers regret any inconvenience this may cause readers, no responsibility for any such changes can be accepted by either the author or the publishers.

Glossary

adapt   change to suit a particular habitat or environment

aerobic respiration   chemical process that takes place in a cell to release energy from sugar using oxygen

agar jelly   gel made from seaweeds

amino acid   building block of proteins

antibiotic   substance that kills or inhibits the growth of bacteria

antibody   special protein made by white blood cells that attaches to any foreign cells and inactivates them

bacteria   single-celled organism that has no nucleus. Bacterial DNA is contained freely in the cytoplasm.

blood transfusion   transfer of blood from one person to another

bone marrow   liquid-like tissue found in the centre of the body's largest bones. Bone marrow is a source of stem cells, and makes red blood cells, some white blood cells, and platelets.

cancer   disease caused by the uncontrollable growth of cells

carbohydrate   chemical, such as sugar or starch, which contains carbon, hydrogen, and oxygen

cartilage   flexible tissue which is found at the surfaces at joints in the skeleton, and also forms structures such as the ear

cell membrane   outer boundary of a cell that surrounds the cytoplasm and controls the movements of molecules into and out of the cell

cellulose   substance found in the cell walls of plants, made of lots of units of sugars joined together in chains

chemotherapy   treatment of cancer using chemicals to kill the cancerous cells

chlorophyll   chemical used by plants to capture the Sun's energy

chloroplast   structure in the plant cell that contains chlorophyll

chromosome   thread-like structure found within the nucleus of cells

clone   genetically identical copy

cloning   making clones

conscious   having an awareness of the surroundings and having sensations and thoughts

cytoplasm   jelly-like substance which fills the cell and in which the components of the cell are suspended

diffuse movement of molecules from an area where they are in high amounts to one where they are in low amounts

DNA (deoxyribose nucleic acid) molecule that carries the genetic code. It is found in the nucleus of the cell.

electron tiny negatively charged particles that travel around the nucleus of an atom

embryo fertilized egg in its early stages of development

enzyme protein molecule that changes the rate of chemical reactions in living things without being affected itself in the process

ethical relating to the issue of right and wrong

fertilization joining together of a male and female sex cell

focus cause light rays to converge on a point

foetus unborn baby more than eight weeks into development

fungi organisms that are neither plants nor animals – they do not move around and cannot photosynthesize

gamete sex cell

gene technology field of technology that involves the manipulation of genetic information

gene unit of inheritance passed on from parent to offspring

glucose type of sugar

heredity handing-down of genes form parent to offspring

hormone chemical message in the body, carried in the blood

immunization process by which a person becomes immune to a disease. Immunization is often carried out by injection of a vaccine.

lens something that focuses rays of light into a sharp image

leukaemia cancer that affects the white blood cells

mitosis type of cell divison where each cell splits to create two identical cells

molecule group of atoms bonded together

nervous system system including the brain, spinal cord, and nerves, that controls the body's responses to things

nucleus central part of the cell, the nucleus contains the DNA

organ part in the body made up of different tissues, carrying out a particular function

organelle specialized structure within a living cell

organism individual living thing, such as a plant or animal

osmosis special type of diffusion that involves the membrane, which is only permeable to certain molecules

permeable allowing the passage of something (for example water or gases)

photosynthesis   process by which green plants make food from carbon dioxide and water, using energy from the Sun

pigment   substance that produces a characteristic colour

plastid   small organelle, found in the cytoplasm of a plant cell, which contains pigment or food

protein   large molecule made from a chain of lots of smaller molecules, called amino acids, joined together

radiation therapy   process by which high-energy radiation from X-rays and other sources are used to kill cancer cells and shrink tumours

resistant   ability of disease-causing organisms to overcome the effect of drugs such as antibiotics

RNA (ribose nucleic acid)   single stranded molecule, found in the nucleus and cytoplasm, which is involved in the production of proteins

starch   carbohydrate made by plants, made up of many glucose molecules joined together in a chain

stem cell   cell that retains the ability to divide and multiply and to create other types of cell

surrogate   female animal that gives birth to another female's baby

tissue   specialized cells grouped together to carry out a function

tuber   undergound stem of a plant

uterus   part of the female reproductive system in which the unborn child grows and develops

vaccinate   to inject a vaccine

vaccine   substance that usually contains weakened or dead bacteria or viruses, injected to bring out an immune response

virus   tiny particle usually made up of DNA or RNA coated in protein. Viruses multiply by infecting other cells.

Index

Titles in the *Life Science in Depth* series include:

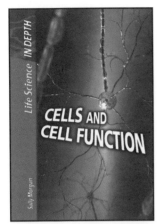

Hardback   0 431 10896 X

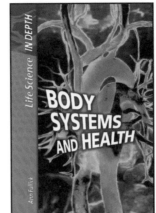

Hardback   0 431 10897 8

Hardback   0 431 10898 6

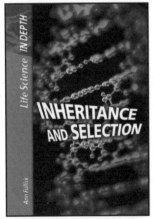

Hardback   0 431 10899 4

Hardback   0 431 10900 1

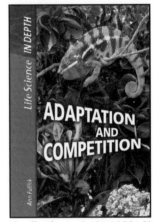

Hardback   0 431 10901 X

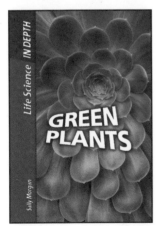

Hardback   0 431 10910 9

Find out about other titles from Heinemann Library on our website www.heinemann.co.uk/library